MATHeMENT

Developing Computational Fluency

James Burnett

ORIGO
EDUCATION

MATHEMENTALS
Developing Computational Fluency

Grade 1
Copyright 2003 ORIGO Education
Author: James Burnett
Project editor: Beth Lewis
Illustrations and design: Brett Cox
Cover design: Brett Cox

For more information, contact:

USA
T   1-888-674-4601
F   1-888-674-4604
E   info@origomath.com
W   www.origomath.com

Canada
E   info@origoeducation.com
W   www.origoeducation.com

ISBN: 978 1 876842 08 6

10 9 8 7 6 5 4 3

CONTENTS

INTRODUCTION

Computational Fluency

The elementary school mathematics curriculum has long focused on the development of formal algorithms to add, subtract, multiply, and divide whole numbers, fractions, and decimals. However, todays students require more than a basic memorization of steps and rules for computation. This is the view of the National Council of Teachers of Mathematics (1) which recently emphasized 'computational fluency' as an essential goal for all students.

The Principles and Standards for School Mathematics described computational fluency as a "connection between conceptual understanding and computational proficiency". Conceptual understanding of computation is knowledge of place value, number relationships, and properties of operations. Computational proficiency is characterized by three main ideas: **efficiency**, **accuracy**, and **flexibility**.

A fluent student will adopt an **efficient** strategy that can be carried out easily without getting bogged down in a series of complex steps. A fluent student will have **accurate** recall of number facts and of important number combinations and show concern for checking results. Computationally fluent students are **flexible** thinkers. They can draw from a range of methods such as mental computation, estimation, calculators, and paper-and-pencil algorithms to solve the problem at hand.

Many resources are available to assist teachers in developing student's skills in estimation, formal written algorithms and in the use of the calculator. There is however a lack of quality materials that assist teachers in developing a range of mental computation strategies. The activities in *Mathementals* are designed to achieve this aim.

The Mathementals Series

Mathementals is a set of six reproducible blackline master books designed to help students develop a range of mental computation strategies. Each book provides appropriate learning experiences with numbers and operations for students working at that level.

The books are divided into sections that focus on one operation. Each section has several *Warm Up* and *Work Out* exercises to provide the guided instruction and written practice. These double-page spreads develop one specific idea within a mental strategy. A range of number models, such as number lines, ten-frames, base-10 blocks, money, and hundred charts, are used throughout these exercises to imitate the thinking required for certain strategies and to give students practice in demonstrating this thinking. *Check Ups* are provided at various points throughout each book. Each of these is followed by a puzzle or game to engage students in mental exercises 'just for fun'. Answers are given at the end of each section.

How to Use this Book

The activities within each book have been developmentally sequenced. This allows teachers to work through the book from the front to the back cover. Alternatively, a confident teacher may want to develop his or her own sequence by drawing on related activities from the different sections of the book. For example, after completing the doubling activities for addition, the teacher may want to focus on the idea of using related doubles to subtract.

The instructional *Warm Up* page can be reproduced as an overhead transparency or as a class handout. The students should be 'walked' through the activity on this page, and encouraged to share their strategies and methods of computation.

The **Mathementals'** character frequently appears with clear, open questions to prompt whole class discussion. There is evidence to suggest that peers can help each other progress from simple to more efficient strategies (2), so teachers should allow plenty of time for students to share their thinking with the class. As there is no one correct strategy, students should be encouraged to use the strategy that works best for them.

After discussing the *Warm Up* page, the students can 'sharpen' their mental skills by independently completing the *Work Out* page. This can be done in school time or as set homework. It is important to correct this page as a class. This will give the students further opportunities to share their strategies and to explain their thinking.

Assessment

It is important to know how confident and competent students are at using a particular strategy. This information can be used to judge whether they need further experiences with that strategy, or if they are ready to progress to the next strategy.

Rubrics are particularly helpful in assessing students' mathematical proficiency in open tasks, such as those that are often used to develop skills in mental computation. The rubric below offers a guide for tracking student progress through the *Warm Up*, *Work Out*, and *Check Up* pages. The findings can be recorded on the student progress summary provided on page 6.

A	The student mentally calculates all examples accurately. The student uses efficient strategies and is able to fully explain his/her thinking and reasoning. The student can describe more than one strategy to solve problems that are alike.
B	The student mentally calculates most examples accurately. The student generally uses efficient strategies and is able to explain his/her thinking and reasoning. The student can sometimes describe more than one strategy to solve problems that are alike.
C	The student mentally calculates some examples accurately. The student uses relatively inefficient strategies and has limited ability to explain his/her thinking and reasoning. The student cannot describe more than one strategy to solve problems that are alike.
D	The student mentally calculates most/all examples inaccurately. The student uses inefficient strategies and has poor/no ability to explain his/her thinking and reasoning. The student cannot describe more than one strategy to solve problems that are alike.

References

1. National Council of Teachers of Mathematics. (2000). *Principles and standards for school mathematics*. Reston, VA: Author.

2. Noddings, N. (1985). Small groups as a setting for research on mathematical problem solving. In E. Silver (Ed.), *Teaching and learning mathematical problem solving: Multiple research perspectives*. Hillsdale, NJ: Lawrence Erlbaum Associates.

Student Progress Summary

Refer to the rubric shown on page 5.
Place a ✔ or write the date in the appropriate column.

			A	B	C	D
ADDITION Strategies	1	Count on 1 – *number facts*				
	2	Count on 2 – *number facts*				
	3	Count on 1 – *beyond the facts*				
	4	Count on 2 – *beyond the facts*				
	5	Count on – *multiples of 10*				
	6	Add the parts – *no regrouping*				
	Check Up 1					
	7	Use doubles – *number facts*				
	8	Use doubles – *multiples of 10*				
	9	Use doubles – *double-plus-1 facts*				
	10	Use doubles – *double-plus-2 facts*				
	11	Make a ten – *number facts*				
	12	Use compatible pairs – *three one-digit addends*				
	Check Up 2					
SUBTRACTION Strategies	13	Count back 1 or 2 – *number facts*				
	14	Count back 1, 2, or 3 – *number facts*				
	15	Count back 1, 2, or 3 – *beyond the facts*				
	16	Count on 1 or 2 – *number facts*				
	17	Use doubles – *number facts*				
	18	Use doubles – *halving multiples of 10*				
	19	Use doubles – *halving with no regrouping*				
	Check Up 3					

ADDITION STRATEGIES

Count on

26 + 2   is   26 ... 27 ... 28.

53 + 30   is   53 ... 63 ... 73 ... 83.

Use doubles

3 + 4   is the same as   double 3 + 1

5 + 7   is the same as   double 5 + 2

Add the parts

26 + 13   is the same as   26 + 10 + 3

Make a ten

9 + 6   is the same as   10 + 5

Use compatible pairs

3 + 9 + 7   is the same as   (3 + 7) + 9

WΩRM UP ➕ 1

Name: _____

Gemma had $5. Her mother gave her $1 more. How much money does she have?

1. Write the number fact.

$$5 + \underline{\hspace{2cm}} = \underline{\hspace{2cm}}$$

How did you figure it out in your head?

2. Look at this count-on card. Complete the number fact.

$$4 + 1 = \underline{\hspace{2cm}}$$

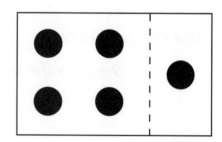

3. Write a number fact for each of these. Write the turnaround fact.

a.

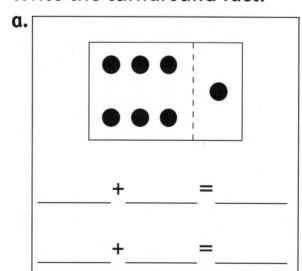

$$\underline{\hspace{1.5cm}} + \underline{\hspace{1.5cm}} = \underline{\hspace{1.5cm}}$$

$$\underline{\hspace{1.5cm}} + \underline{\hspace{1.5cm}} = \underline{\hspace{1.5cm}}$$

b.

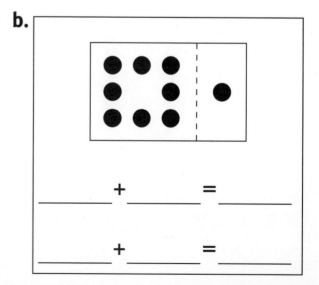

$$\underline{\hspace{1.5cm}} + \underline{\hspace{1.5cm}} = \underline{\hspace{1.5cm}}$$

$$\underline{\hspace{1.5cm}} + \underline{\hspace{1.5cm}} = \underline{\hspace{1.5cm}}$$

Name: _____

1. Write a number fact to show each total.

a.
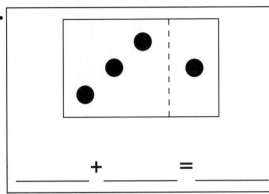

_____ + _____ = _____

b.
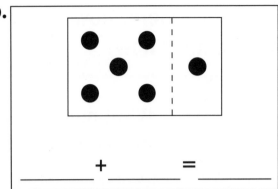

_____ + _____ = _____

c.
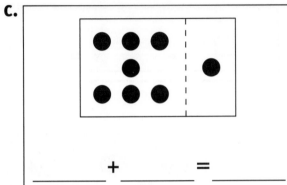

_____ + _____ = _____

d.
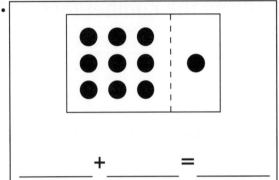

_____ + _____ = _____

2. Write the number fact then write the turnaround fact.

a.
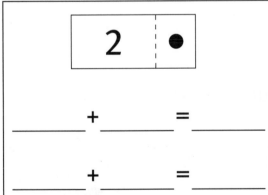

_____ + _____ = _____

_____ + _____ = _____

b.
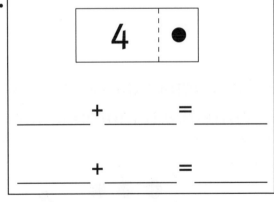

_____ + _____ = _____

_____ + _____ = _____

c.
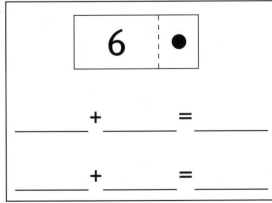

_____ + _____ = _____

_____ + _____ = _____

d.

_____ + _____ = _____

_____ + _____ = _____

WARM UP  2

Name: _____

Jack had 6 trading cards. His friend gave him 2 more. How many cards does he have now?

 **What number did you put in your head first?**

1. Write the number fact.

_6_ _____ + _____ = _____

2. Look at this count-on card. Complete the number fact.

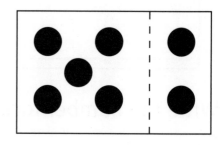

_5_ _____ + _2_ _____ = _____

3. For each of these, write the number fact. Write the turnaround fact.

a.

_____ + _____ = _____

_____ + _____ = _____

b.

_____ + _____ = _____

_____ + _____ = _____

Name: _____

1. Write a number fact to show each total.

a.
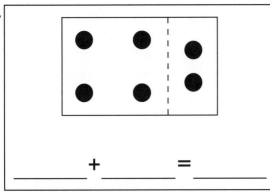
___ + ___ = ___

b.
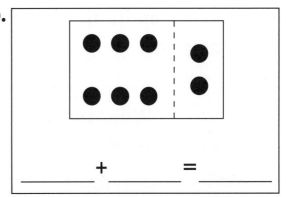
___ + ___ = ___

c.
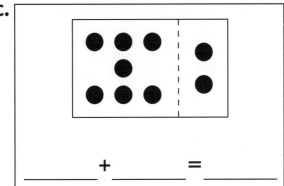
___ + ___ = ___

d.
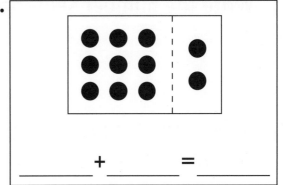
___ + ___ = ___

2. Write the number fact then write the turnaround fact.

a.

| 5 | •• |

___ + ___ = ___

___ + ___ = ___

b.
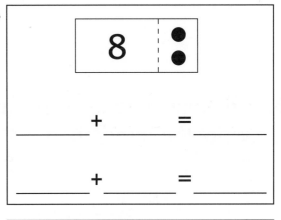

| 8 | •• |

___ + ___ = ___

___ + ___ = ___

c.
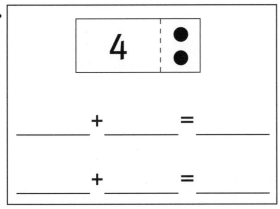

| 4 | •• |

___ + ___ = ___

___ + ___ = ___

d.

| 6 | •• |

___ + ___ = ___

___ + ___ = ___

WaRM UP  3

Name: _____

There were 17 swimmers in the pool.
One more dived in. How many swimmers
are in the pool?

 **What number did you put
in your head first?**

1. Write the number sentence.

    17 + _____ = _____

2. There are 15 cubes in this cup.
    One more cube is added.
    Complete the number sentence.

    15 + 1 = _____

3. Add one more to each cup. Write the number sentence.
    Write the turnaround.

    a.

    18

    _____ + _____ = _____

    _____ + _____ = _____

    b.

    12

    _____ + _____ = _____

    _____ + _____ = _____

Name: _____

1. Write two number sentences for each of these.

a.

16 + 1 = _____

1 + 16 = _____

b.

_____ + _____ = _____

_____ + _____ = _____

c.

_____ + _____ = _____

_____ + _____ = _____

d.

_____ + _____ = _____

_____ + _____ = _____

2. Write the answer. Ring the number you put in your head first.

a. 18 + 1 = _____   **b.** 1 + 20 = _____   **c.** 13 + 1 = _____

d. 11 + 1 = _____   **e.** 1 + 22 = _____   **f.** 1 + 16 = _____

3. Add $1 to each of these prices. Write the new price.

a.
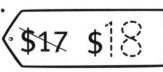
$17 $18

b.
$25

c.
$30

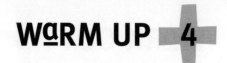

Name: _____

**Joey saved $16. He was given $2 more.
How much money does he have now?**

1. Write the number sentence.

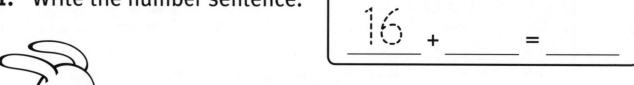

16 + _____ = _____

**What number did you put
in your head first?**

2. There is $23 in the piggy bank.
Add $2 more.
Complete the number sentence.

$23

23 + 2 = _____

3. Add $2 to each piggy bank.
Write two different number sentences to show the total.

a.

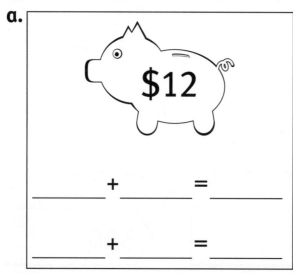

$12

_____ + _____ = _____

_____ + _____ = _____

b.

$15

_____ + _____ = _____

_____ + _____ = _____

Name: _____

1. Add $2 to each amount. Write two number sentences.

a.

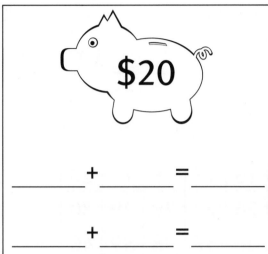

_____ + _____ = _____

_____ + _____ = _____

b.

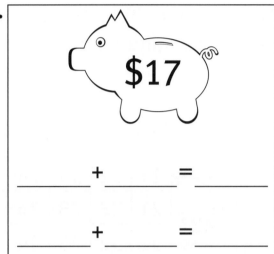

_____ + _____ = _____

_____ + _____ = _____

c.

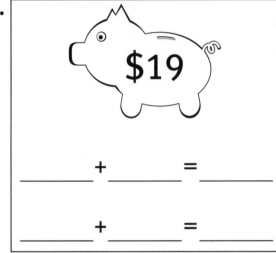

_____ + _____ = _____

_____ + _____ = _____

d.

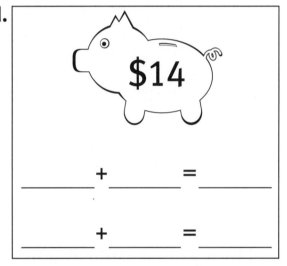

_____ + _____ = _____

_____ + _____ = _____

2. Color a number then count on 2.
Write a matching number sentence.

a.

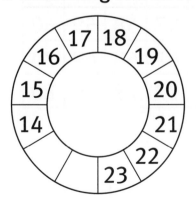

b.

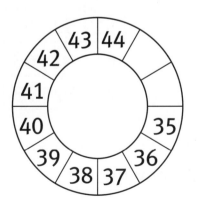

_____ + _____ = _____   _____ + _____ = _____

 **WaRM UP** 5

Name: _____

**It took Calvin 26 minutes to walk to
the store and 30 minutes to walk back.
How many minutes did he walk?**

1. a. Use this counting board to help you figure out the total.

11	12	13	14	15	16	17	18	19	20
21	22	23	24	25	26	27	28	29	30
31	32	33	34	35	36	37	38	39	40
41	42	43	44	45	46	47	48	49	50
51	52	53	54	55	56	57	58	59	60

b. Draw arrows on the board to show what you did.

c. Write a matching
number sentence.

_____ + _____ = _____

2. Look at this counting board.

31	32	33	34	35	36	37	38	39	40
41	42	43	44	45	46	47	48	49	50
51	52	53	54	55	56	57	58	59	60
61	62	63	64	65	66	67	68	69	70
71	72	73	74	75	76	77	78	79	80

Write a number sentence to
show what has happened.

_____ + _____ = _____

Name: _____

1. Write the answers to these.
    Draw arrows on the hundred chart to show what you did.

    a. 35 + 20 = _____   **b.** 43 + 30 = _____   **c.** 51 + 40 = _____

    d. 42 + 20 = _____   **e.** 56 + 30 = _____   **f.** 38 + 40 = _____

1	2	3	4	5	6	7	8	9	10
11	12	13	14	15	16	17	18	19	20
21	22	23	24	25	26	27	28	29	30
31	32	33	34	35	36	37	38	39	40
41	42	43	44	45	46	47	48	49	50
51	52	53	54	55	56	57	58	59	60
61	62	63	64	65	66	67	68	69	70
71	72	73	74	75	76	77	78	79	80
81	82	83	84	85	86	87	88	89	90
91	92	93	94	95	96	97	98	99	100

2. Write the number that belongs at the end of each piece.

    a. 33

    b. 78

    c. 52

    d. 29

WaRM UP  6

Name: _____

1. Look at this counting board.

    a. Place a counter on 25. Move it forward 21 spaces.

1	2	3	4	5	6	7	8	9	10
11	12	13	14	15	16	17	18	19	20
21	22	23	24	(25)	26	27	28	29	30
31	32	33	34	35	36	37	38	39	40
41	42	43	44	45	46	47	48	49	50

    b. Draw arrows on the board to show how you added 21.

    c. Write a number sentence to match what you did.

    25 + _____

2. Look at this counting board.

1	2	3	4	5	6	7	8	9	10
11	12	13	14	15	16	17	18	19	20
21	22	23	24	25	26	(27)	28	29	30
31	32	33	34	35	36	37	38	39	40
41	42	43	44	45	46	47	48	49	50

    a. Complete this number sentence to show what has happened.

    27 + _____

    b. Draw more arrows to show another way to add 12.

Add the Parts

1. Draw arrows on the hundred chart below to show how you could add each of these. Write the answer.

    a. 22 + 32 = _____   **b.** 41 + 33 = _____   **c.** 29 + 21 = _____

    d. 55 + 32 = _____   **e.** 66 + 12 = _____   **f.** 17 + 31 = _____

1	2	3	4	5	6	7	8	9	10
11	12	13	14	15	16	17	18	19	20
21	22	23	24	25	26	27	28	29	30
31	32	33	34	35	36	37	38	39	40
41	42	43	44	45	46	47	48	49	50
51	52	53	54	55	56	57	58	59	60
61	62	63	64	65	66	67	68	69	70
71	72	73	74	75	76	77	78	79	80
81	82	83	84	85	86	87	88	89	90
91	92	93	94	95	96	97	98	99	100

2. Write the number that belongs at the end of each piece.

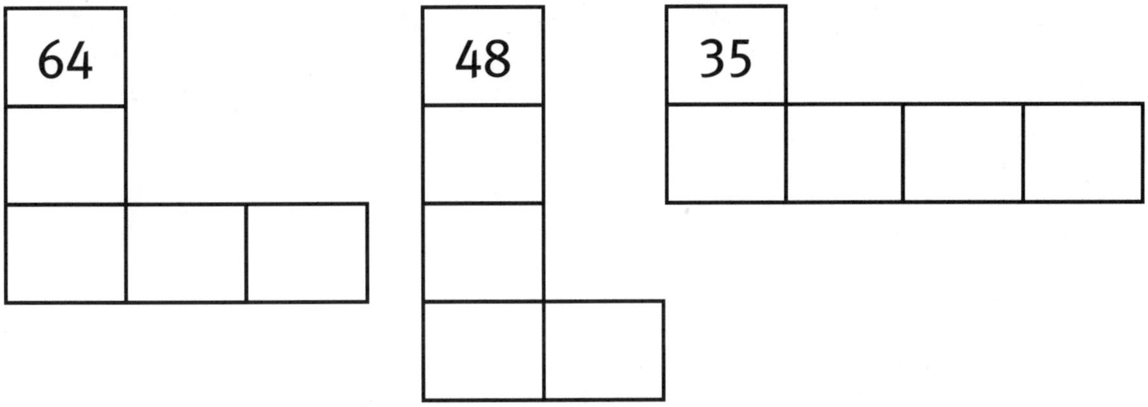

64

48

35

CHECK UP 1

Name: _____

1. Figure out these in your head. Write the answers.

   a. $8 + 1 =$ _____   **b.** $17 + 1 =$ _____   **c.** $7 + 2 =$ _____

   d. $23 + 2 =$ _____   **e.** $23 + 20 =$ _____   **f.** $35 + 22 =$ _____

2. **a.** Write the answer.

   $43 + 21 =$ _____

   Think about how you figured it out.

   b. Write two number sentences you could solve the same way. Make the answers less than 100.

   _____ + _____ = _____   _____ + _____ = _____

3. Write the answers in the baskets.

   a.  34 + 20

   b. 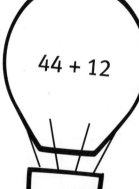 44 + 12

   c. 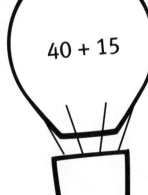 40 + 15

   d. 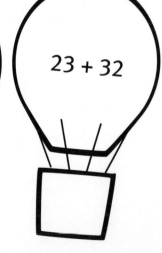 23 + 32

Name: _____

Figure out each of these and write the answer.
Find the part below that has the answer and color it to match.

a. 23 + 2 = _____ (yellow)   **b.** 28 + 1 = _____ (light green)

c. 38 + 20 = _____ (orange)   **d.** 46 + 21 = _____ (dark blue)

e. 16 + 2 = _____ (purple)   **f.** 45 + 40 = _____ (dark green)

g. 32 + 1 = _____ (red)   **h.** 20 + 32 = _____ (light blue)

WARM UP +7

Name: _____

1. Draw dots on this domino to show double 9.

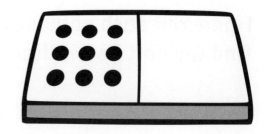

Write the number fact.

_____ + _____ = _____

How did you figure out double 9 in your head?

2. For each of these, draw dots to show a double. Write the double fact.

a.
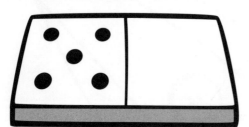

_____ + _____ = _____

b.
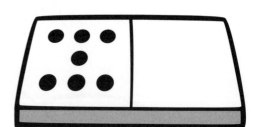

_____ + _____ = _____

3. Draw dots to show the doubles that have these answers. Complete the number facts.

a.
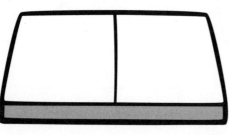

_____ + _____ = 16

b.
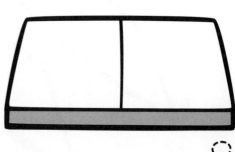

_____ + _____ = 8

Use Doubles

Name: _____

1. Complete the double fact for each of these.

a.
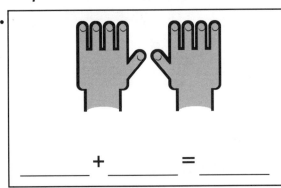
_____ + _____ = _____

b.
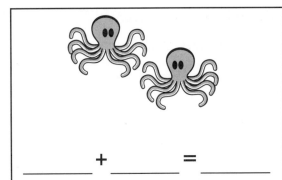
_____ + _____ = _____

c.

_____ + _____ = _____

d.
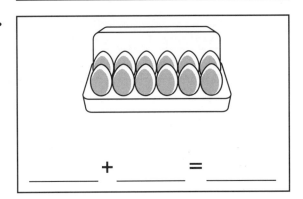
_____ + _____ = _____

2. Write the double facts that have these answers.

a. 14 = _____ + _____

b. 20 = _____ + _____

c. 4 = _____ + _____

d. 18 = _____ + _____

3. Draw dots on these dominoes to show other doubles. Write the facts.

a.
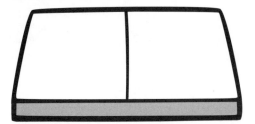
_____ + _____ = _____

b.

_____ + _____ = _____

WARM UP 8

This barbell has 20 kilograms on each end. How many kilograms are on the bar?

How can you use double 2 to help figure out the answer?

1. Complete this sentence.

    > Double 2 is _____ **so** Double 20 is _____

2. Complete a sentence for each barbell.

    a.
    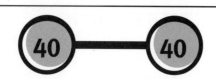

    Double 4 = _____ **so**

    Double 40 = _____

    b.

    Double 1 = _____ **so**

    Double 10 = _____

    c.

    Double 3 = _____ **so**

    Double 30 = _____

    d.
    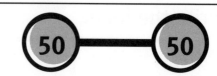

    Double 5 = _____ **so**

    Double 50 = _____

Use Doubles

Name: ______________________

WORK OUT 8

1. Draw an arrow to the double you could use to help figure out the total. Write the answer.

a.	Double 30 = ______	Double 4 = 8
b.	Double 50 = ______	Double 2 = 4
c.	Double 20 = ______	Double 3 = 6
d.	Double 40 = ______	Double 5 = 10

2. Double each of these. Write the number sentence.

a. 30 ______ + ______ = ______

b. 10 ______ + ______ = ______

c. 50 ______ + ______ = ______

3. Write numbers in each of these to make them true. Make the numbers in the circles less than 10.

a. Double ◯ = ______ **so** double ______ = ______

b. Double ◯ = ______ **so** double ______ = ______

WARM UP 9

Name: _____

1. Complete an addition fact for each domino.

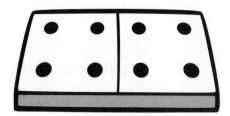   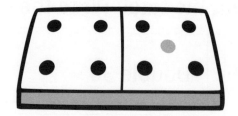

_____ + _____ = _____   _____ + _____ = _____

How can you use a double to figure out 4 + 5?

2. Write the double fact.
Draw one more dot then write the 'near double' fact.

a.
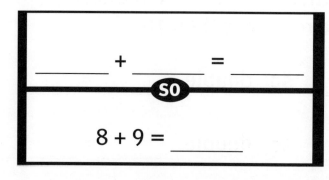

2 + 2 = _____

2 + _____ = _____

b.
6 + _____ = _____

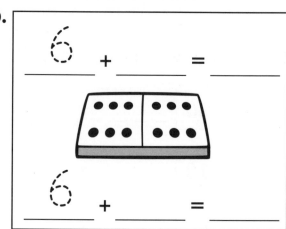

6 + _____ = _____

3. Write a double you could use to help figure out 8 + 9.
Write the answer.

_____ + _____ = _____

SO

8 + 9 = _____

What other double could you use?

Use Doubles

Name: _____

1. Write the double fact. Draw one more dot then write two 'near double' facts.

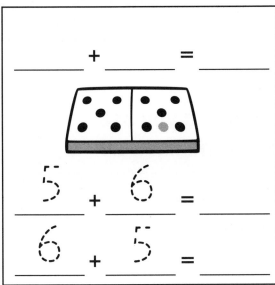

a.
_____ + _____ = _____

$5 + 6 =$ _____

$6 + 5 =$ _____

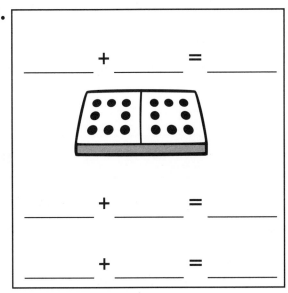

b.
_____ + _____ = _____

_____ + _____ = _____

_____ + _____ = _____

2. Complete each of these.

a.
3 + 3 = _____

SO

3 + 4 = _____

b.
7 + 7 = _____

SO

7 + 8 = _____

c.
2 + 2 = _____

SO

2 + 3 = _____

3. Write the answer then write the turnaround fact.

a. 5 + 6 = ⟨_____⟩ = _____ + _____

b. 3 + 4 = ⟨_____⟩ = _____ + _____

c. 7 + 8 = ⟨_____⟩ = _____ + _____

d. 4 + 5 = ⟨_____⟩ = _____ + _____

Name: _____

1. Complete the matching number fact.

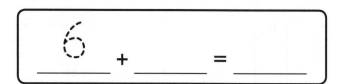

6
_____ + _____ = _____

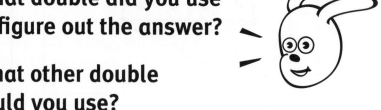

What double did you use to figure out the answer?

What other double could you use?

2. Write the double fact.
Draw two more dots then write the 'near double' fact.

a.

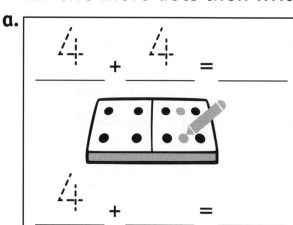

4 + 4 = _____
_____

4 + _____ = _____

b.

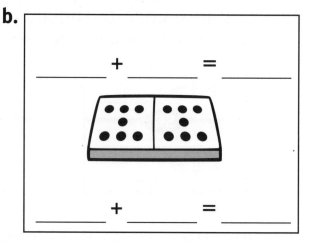

_____ + _____ = _____

_____ + _____ = _____

3. Complete a fact to match each picture.

a.

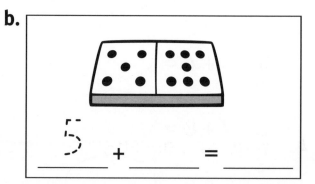

3 + _____ = _____

b.

5 + _____ = _____

Name: _____

1. Write the double fact. Draw two more dots then write two 'near double' facts.

a.
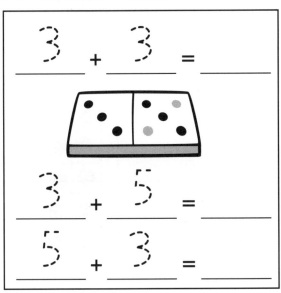

_____ 3 + 3 _____ = _____

_____ 3 + 5 _____ = _____

_____ 5 + 3 _____ = _____

b.
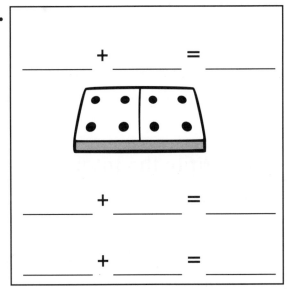

_____ + _____ = _____

_____ + _____ = _____

_____ + _____ = _____

2. Complete each of these.

a.
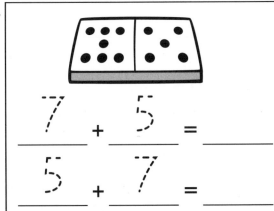

_____ 7 + 5 _____ = _____

_____ 5 + 7 _____ = _____

b.
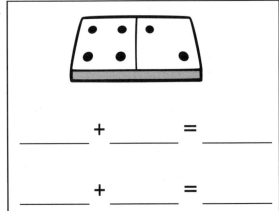

_____ + _____ = _____

_____ + _____ = _____

3. Complete these 'near double' facts.

a. _____ + 5 = 8   b. 6 + _____ = 14   c. 4 + 6 = _____

d. _____ + 4 = 6   e. 7 + _____ = 16   f. 5 + 7 = _____

4. Write the missing numbers.

a. 8 + 6 = _____ = 6 + 8   b. 10 + 8 = _____ = 8 + 10

Name: _____

**Jade had 9 girls and 6 boys at her party.
How many guests in all?**

1. **a.** These ten-frames show 9 counters.
    Draw 6 more counters.

    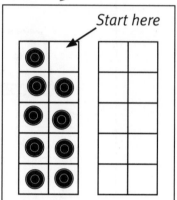

    Start here

    b. Write the total. _____

 **What did you notice?**

   c. Complete the sentence.

   > 9 + 6 **is the same as** 10 + _____

2. **a.** These ten-frames show 8.
    Draw 4 more counters.

    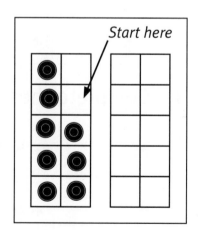

    Start here

    b. Write the total. _____

    c. Complete the sentence.

    > 8 + 4 **is the same as** 10 + _____

Make a Ten

Name: _____

1. Draw more counters then complete the sentence.

a.

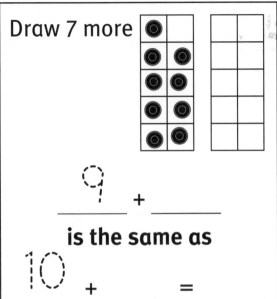

Draw 7 more

_9_ + _____

is the same as

_10_ + _____ = _____

b.

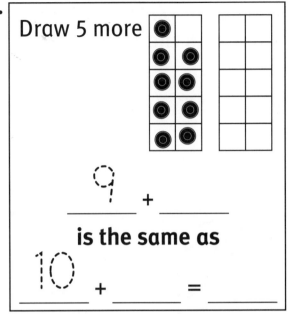

Draw 5 more

_9_ + _____

is the same as

_10_ + _____ = _____

c.

Draw 5 more

_____ + _____

is the same as

_____ + _____ = _____

d.

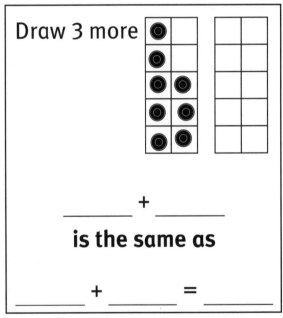

Draw 3 more

_____ + _____

is the same as

_____ + _____ = _____

2. For each of these, draw an arrow to a number sentence below that has the same answer. Write the answer.

a. | 9 + 8 |

b. | 8 + 6 |

c. | 9 + 3 |

| 10 + 2 = _____ |

| 10 + 7 = _____ |

| 10 + 4 = _____ |

WⱭRM UP 12

Name: _____

Rosie bought three books.
They cost $7, $6, and $4.
What was the total cost?

1. **a.** Write the numbers.

$$\_\_\_\_\_ + \_\_\_\_\_ + \_\_\_\_\_$$

 **Look for an easy way to figure out the answer.**

b. Ring two numbers that add to make ten.

c. Write the total. _____

2. Suppose the books were $3, $5, and $7.

Find a pair of numbers that add nicely together.

Write the number sentence.

$$\_\_\_\_\_ + \_\_\_\_\_ + \_\_\_\_\_ = \_\_\_\_\_$$

Use Compatible Pairs

Name: _____

1. For each of these, ring the two numbers you would add first.
Write the total.

a.
```
7       3
    4
```
Total _____

b.
```
        7
3
        6
```
Total _____

c.
```
            2
    8
        7
```
Total _____

d.
```
7       6
    4
```
Total _____

e.
```
        9
1
        7
```
Total _____

f.
```
            6
    5
            5
```
Total _____

2. Figure out the total score for each of these targets.

a.
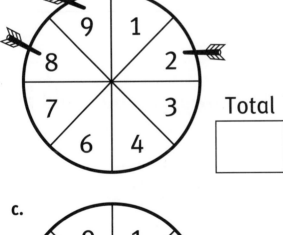
Total ▢

b.
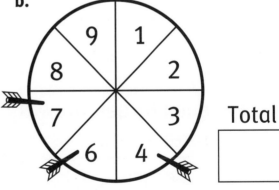
Total ▢

c.
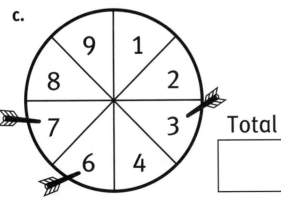
Total ▢

d.
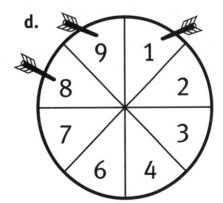
Total ▢

Use Compatible Pairs   33

CH**e**CK UP  2

Name: _____

1. Figure out these in your head. Write the answers.

    a. $6 + 6 =$ _____
    b. $30 + 30 =$ _____
    c. $4 + 5 =$ _____

    d. $6 + 8 =$ _____
    e. $9 + 5 =$ _____
    f. $7 + 5 + 3 =$ _____

2. Write the answer.

    | $7 + 9 =$ _____ |
    Explain two ways you could figure this out.

3. Write two near doubles you can solve using this double.

    _____ + _____ = _____   **3 + 3**   _____ + _____ = _____

4. For each target, draw three ➶ to show how you could make the total score.

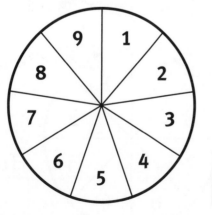

Total
19

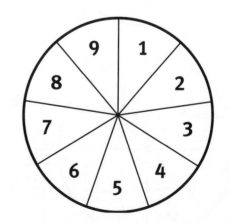

Total
13

JUST FOR FUN +2

Figure out each of these and write the answer.
Find the part below that has the answer and color it to match.

a. 9 + 9 = _____ (light blue)   **b.** 40 + 40 = _____ (red)

c. 7 + 8 = _____ (orange)   **d.** 3 + 5 = _____ (dark blue)

e. 20 + 20 = _____ (purple)   **f.** 9 + 5 = _____ (dark green)

g. 8 + 3 = _____ (light green)   **h.** 5 + 6 + 5 = _____ (yellow)

Name: _____

Gemma had $5. Her mother gave her $1 more. How much money does she have?

1. Write the number fact.

$$5 + \$1 = \$6$$

How did you figure it out in your head?

2. Look at this count-on card. Complete the number fact.

$$4 + 1 = 5$$

3. Write a number fact for each of these. Write the turnaround fact.

a.
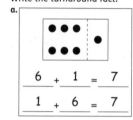
$$6 + 1 = 7$$
$$1 + 6 = 7$$

b.
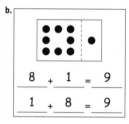
$$8 + 1 = 9$$
$$1 + 8 = 9$$

Name: _____

 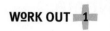

1. Write a number fact to show each total.

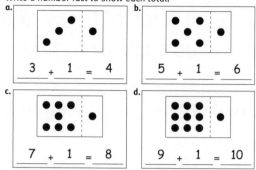

a.
$$3 + 1 = 4$$

b.
$$5 + 1 = 6$$

c.
$$7 + 1 = 8$$

d.
$$9 + 1 = 10$$

2. Write the number fact then write the turnaround fact.

a.
| 2 | • |

$$2 + 1 = 3$$
$$1 + 2 = 3$$

b.
| 4 | • |

$$4 + 1 = 5$$
$$1 + 4 = 5$$

c.
| 6 | • |

$$6 + 1 = 7$$
$$1 + 6 = 7$$

d.
| 8 | • |

$$8 + 1 = 9$$
$$1 + 8 = 9$$

Name: _____

Jack had 6 trading cards. His friend gave him 2 more. How many cards does he have now?

What number did you put in your head first?

1. Write the number fact.

$$6 + 2 = 8$$

2. Look at this count-on card. Complete the number fact.

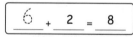

$$5 + 2 = 7$$

3. For each of these, write the number fact. Write the turnaround fact.

a.
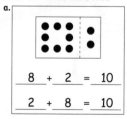
$$8 + 2 = 10$$
$$2 + 8 = 10$$

b.
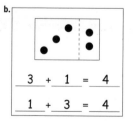
$$3 + 1 = 4$$
$$1 + 3 = 4$$

Name: _____

1. Write a number fact to show each total.

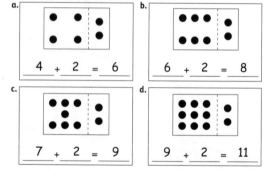

a.
$$4 + 2 = 6$$

b.
$$6 + 2 = 8$$

c.
$$7 + 2 = 9$$

d.
$$9 + 2 = 11$$

2. Write the number fact then write the turnaround fact.

a.
| 5 | •• |

$$5 + 2 = 7$$
$$2 + 5 = 7$$

b.
| 8 | •• |

$$8 + 2 = 10$$
$$2 + 8 = 10$$

c.
| 4 | •• |

$$4 + 2 = 6$$
$$2 + 4 = 6$$

d.
| 6 | •• |

$$6 + 2 = 8$$
$$2 + 6 = 8$$

WARM UP 3

Name: _____

There were 17 swimmers in the pool. One more dived in. How many swimmers are in the pool?

 What number did you put in your head first?

1. Write the number sentence.

$\underline{17} + \underline{1} = \underline{18}$

2. There are 15 cubes in this cup. One more cube is added. Complete the number sentence.

15

$\underline{15} + 1 = \underline{16}$

3. Add one more to each cup. Write the number sentence. Write the turnaround.

a.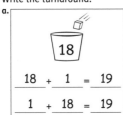
18

$\underline{18} + \underline{1} = \underline{19}$
$\underline{1} + \underline{18} = \underline{19}$

b.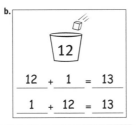
12

$\underline{12} + \underline{1} = \underline{13}$
$\underline{1} + \underline{12} = \underline{13}$

12   Count On

WORK OUT 3

Name: _____

1. Write two number sentences for each of these.

a.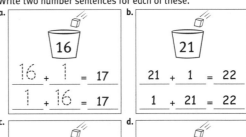
16
$\underline{16} + \underline{1} = 17$
$\underline{1} + \underline{16} = 17$

b.
21
$\underline{21} + \underline{1} = 22$
$\underline{1} + \underline{21} = 22$

c.
14
$\underline{14} + \underline{1} = 15$
$\underline{1} + \underline{14} = 15$

d.
19
$\underline{19} + \underline{1} = 20$
$\underline{1} + \underline{19} = 20$

2. Write the answer. Ring the number you put in your head first.

a. ⟨18⟩ + 1 = $\underline{19}$ b. 1 + ⟨20⟩ = $\underline{21}$ c. ⟨13⟩ + 1 = $\underline{14}$

d. ⟨11⟩ + 1 = $\underline{12}$ e. 1 + ⟨22⟩ = $\underline{23}$ f. 1 + ⟨16⟩ = $\underline{17}$

3. Add $1 to each of these prices. Write the new price.

a. ⟨$17 $18⟩   b. ⟨$25 $26⟩   c. ⟨$30 $31⟩

Count On   13

WARM UP 4

Name: _____

Joey saved $16. He was given $2 more. How much money does he have now?

1. Write the number sentence.

$\underline{16} + \underline{\$2} = \underline{\$18}$

 What number did you put in your head first?

2. There is $23 in the piggy bank. Add $2 more. Complete the number sentence.

$23

$\underline{23} + \underline{2} = \$25$

3. Add $2 to each piggy bank. Write two different number sentences to show the total.

a.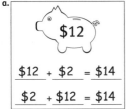
$12
$\underline{\$12} + \underline{\$2} = \underline{\$14}$
$\underline{\$2} + \underline{\$12} = \underline{\$14}$

b.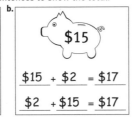
$15
$\underline{\$15} + \underline{\$2} = \underline{\$17}$
$\underline{\$2} + \underline{\$15} = \underline{\$17}$

14   Count On

WORK OUT 4

Name: _____

1. Add $2 to each amount. Write two number sentences.

a.
$20
$\$20 + \$2 = \$22$
$\$2 + \$20 = \$22$

b.
$17
$\$17 + \$2 = \$19$
$\$2 + \$17 = \$19$

c.
$19
$\$19 + \$2 = \$21$
$\$2 + \$19 = \$21$

d.
$14
$\$14 + \$2 = \$16$
$\$2 + \$14 = \$16$

2. Color a number then count on 2. Write a matching number sentence.

a. 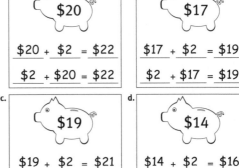 ★   b. ★

$\underline{17} + \underline{2} = \underline{19}$ $\underline{41} + \underline{2} = \underline{43}$

Count On   15

★ Answers will vary. This is one example.

WARM UP 5

Name: _____

It took Calvin 26 minutes to walk to the store and 30 minutes to walk back. How many minutes did he walk?

1. a. Use this counting board to help you figure out the total.

11	12	13	14	15	16	17	18	19	20
21	22	23	24	25	26	27	28	29	30
31	32	33	34	35	36	37	38	39	40
41	42	43	44	45	46	47	48	49	50
51	52	53	54	55	56	57	58	59	60

   b. Draw arrows on the board to show what you did.

   c. Write a matching number sentence.

   $\underline{26} + \underline{30} = \underline{56}$

2. Look at this counting board.

31	32	33	34	35	36	37	38	39	40
41	42	43	44	45	46	47	48	49	50
51	52	53	54	55	56	57	58	59	60
61	62	63	64	65	66	67	68	69	70
71	72	73	74	75	76	77	78	79	80

   Write a number sentence to show what has happened.

   $\underline{32} + \underline{40} = \underline{72}$

Name: _____

WORK OUT 5

1. Write the answers to these.
   Draw arrows on the hundred chart to show what you did.

   a. $35 + 20 = \underline{55}$ b. $43 + 30 = \underline{73}$ c. $51 + 40 = \underline{91}$

   d. $42 + 20 = \underline{62}$ e. $56 + 30 = \underline{86}$ f. $38 + 40 = \underline{78}$

1	2	3	4	5	6	7	8	9	10
11	12	13	14	15	16	17	18	19	20
21	22	23	24	25	26	27	28	29	30
31	32	33	34	35	36	37	38	39	40
41	42	43	44	45	46	47	48	49	50
51	52	53	54	55	56	57	58	59	60
61	62	63	64	65	66	67	68	69	70
71	72	73	74	75	76	77	78	79	80
81	82	83	84	85	86	87	88	89	90
91	92	93	94	95	96	97	98	99	100

2. Write the number that belongs at the end of each piece.

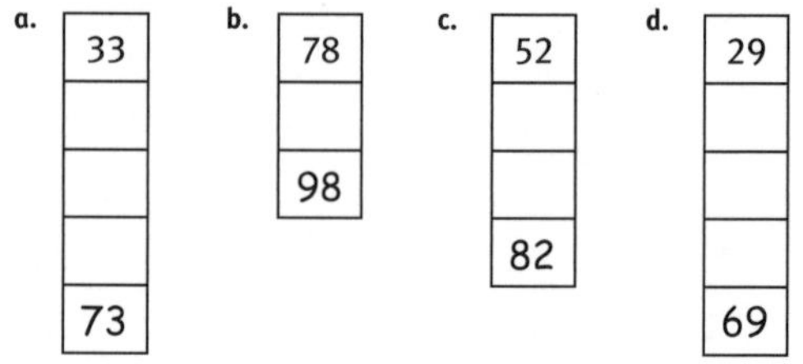

   a.

33
73

   b.

78
98

   c.

52
82

   d.

29
69

WARM UP 6

Name: _____

1. Look at this counting board.
   a. Place a counter on 25. Move it forward 21 spaces.

1	2	3	4	5	6	7	8	9	10
11	12	13	14	15	16	17	18	19	20
21	22	23	24	25	26	27	28	29	30
31	32	33	34	35	36	37	38	39	40
41	42	43	44	45	46	47	48	49	50

   b. Draw arrows on the board to show how you added 21.

   c. Write a number sentence to match what you did.

   $25 + \underline{20 + 1 = 46}$

2. Look at this counting board.

1	2	3	4	5	6	7	8	9	10
11	12	13	14	15	16	17	18	19	20
21	22	23	24	25	26	27	28	29	30
31	32	33	34	35	36	37	38	39	40
41	42	43	44	45	46	47	48	49	50

   a. Complete this number sentence to show what has happened.

   $27 + \underline{10 + 2 = 39}$

   b. Draw more arrows to show another way to add 12. *

Name: _____

WORK OUT 6

1. Draw arrows on the hundred chart below to show how you could add each of these. Write the answer.

   a. $22 + 32 = \underline{54}$ b. $41 + 33 = \underline{74}$ c. $29 + 21 = \underline{50}$

   d. $55 + 32 = \underline{87}$ e. $66 + 12 = \underline{78}$ f. $17 + 31 = \underline{48}$

1	2	3	4	5	6	7	8	9	10
11	12	13	14	15	16	17	18	19	20
21	22	23	24	25	26	27	28	29	30
31	32	33	34	35	36	37	38	39	40
41	42	43	44	45	46	47	48	49	50
51	52	53	54	55	56	57	58	59	60
61	62	63	64	65	66	67	68	69	70
71	72	73	74	75	76	77	78	79	80
81	82	83	84	85	86	87	88	89	90
91	92	93	94	95	96	97	98	99	100

*

2. Write the number that belongs at the end of each piece.

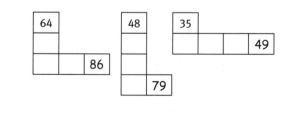

★ Answers will vary. This is one example.

CHECK UP 1

Name: _____

1. Figure out these in your head. Write the answers.

   a. 8 + 1 = __9__   b. 17 + 1 = __18__   c. 7 + 2 = __9__

   d. 23 + 2 = __25__   e. 23 + 20 = __43__   f. 35 + 22 = __57__

2. a. Write the answer.

   43 + 21 = __64__

   Think about how you figured it out.

   b. Write two number sentences you could solve the same way. Make the answers less than 100.

   | __32__ + __43__ = __75__ | ★ | __15__ + __63__ = __78__ | ★ |

3. Write the answers in the baskets.

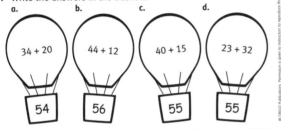

   a. 34 + 20 → 54   b. 44 + 12 → 56   c. 40 + 15 → 55   d. 23 + 32 → 55

20   Check Up

Name: _____   JUST FOR FUN 1

Figure out each of these and write the answer.
Find the part below that has the answer and color it to match.

   a. 23 + 2 = __25__ (yellow)   b. 28 + 1 = __29__ (light green)

   c. 38 + 20 = __58__ (orange)   d. 46 + 21 = __67__ (dark blue)

   e. 16 + 2 = __18__ (purple)   f. 45 + 40 = __85__ (dark green)

   g. 32 + 1 = __33__ (red)   h. 20 + 32 = __52__ (light blue)

Just for Fun   21

WARM UP 7

Name: _____

1. Draw dots on this domino to show double 9.

   Write the number fact.

   __9__ + __9__ = __18__

   How did you figure out double 9 in your head?

2. For each of these, draw dots to show a double. Write the double fact.

   a. 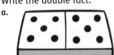   b.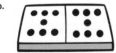

   __5__ + __5__ = __10__   __7__ + __7__ = __14__

3. Draw dots to show the doubles that have these answers. Complete the number facts.

   a. 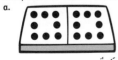   b.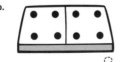

   __8__ + __8__ = 16   __4__ + __4__ = 8

22   Use Doubles

Name: _____   WORK OUT 7

1. Complete the double fact for each of these.

   a.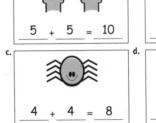
   __5__ + __5__ = __10__

   b. __8__ + __8__ = __16__

   c. __4__ + __4__ = __8__

   d.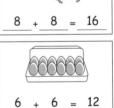
   __6__ + __6__ = __12__

2. Write the double facts that have these answers.

   a. 14 = __7__ + __7__   b. 20 = __10__ + __10__

   c. 4 = __2__ + __2__   d. 18 = __9__ + __9__

3. Draw dots on these dominoes to show other doubles. Write the facts.

   a. 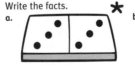 ★   b. 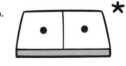 ★

   __3__ + __3__ = __6__   __1__ + __1__ = __2__

Use Doubles   23

★ Answers will vary. This is one example.

Answers   39

Name: _____

This barbell has 20 kilograms on each end. How many kilograms are on the bar?

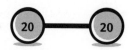

 How can you use double 2 to help figure out the answer?

1. Complete this sentence.

Double 2 is __4__ so Double 20 is __40__

2. Complete a sentence for each barbell.

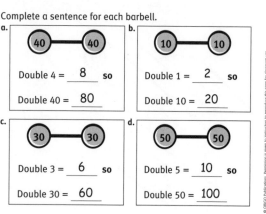

a.
Double 4 = __8__ so
Double 40 = __80__

b.
Double 1 = __2__ so
Double 10 = __20__

c.
Double 3 = __6__ so
Double 30 = __60__

d.
Double 5 = __10__ so
Double 50 = __100__

24   Use Doubles

Name: _____

1. Draw an arrow to the double you could use to help figure out the total. Write the answer.

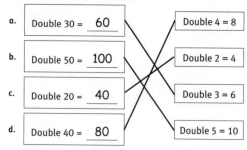

a. Double 30 = __60__
b. Double 50 = __100__
c. Double 20 = __40__
d. Double 40 = __80__

Double 4 = 8
Double 2 = 4
Double 3 = 6
Double 5 = 10

2. Double each of these. Write the number sentence.

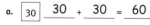

a. | 30 | __30__ + __30__ = __60__
b. | 10 | __10__ + __10__ = __20__
c. | 50 | __50__ + __50__ = __100__

3. Write numbers in each of these to make them true. Make the numbers in the circles less than 10. ★

a. Double (9) = __18__ so double __90__ = __180__

b. Double (7) = __14__ so double __70__ = __140__

Use Doubles   25

 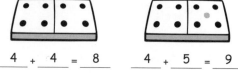

Name: _____

1. Complete an addition fact for each domino.

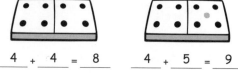

__4__ + __4__ = __8__   __4__ + __5__ = __9__

 How can you use a double to figure out 4 + 5?

2. Write the double fact.
Draw one more dot then write the 'near double' fact.

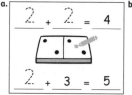

a.
__2__ + __2__ = __4__
__2__ + __3__ = __5__

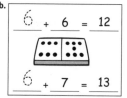

b.
__6__ + __6__ = __12__
__6__ + __7__ = __13__

3. Write a double you could use to help figure out 8 + 9. Write the answer.

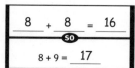

__8__ + __8__ = __16__
SO
8 + 9 = __17__

 What other double could you use?

26   Use Doubles

Name: _____

 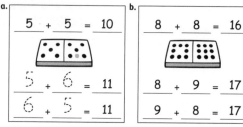

1. Write the double fact. Draw one more dot then write two 'near double' facts.

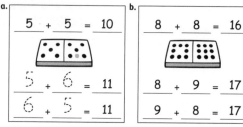

a.
__5__ + __5__ = __10__
__5__ + __6__ = __11__
__6__ + __5__ = __11__

b.
__8__ + __8__ = __16__
__8__ + __9__ = __17__
__9__ + __8__ = __17__

2. Complete each of these.

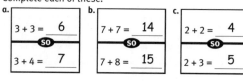

a.
3 + 3 = __6__
SO
3 + 4 = __7__

b.
7 + 7 = __14__
SO
7 + 8 = __15__

c.
2 + 2 = __4__
SO
2 + 3 = __5__

3. Write the answer then write the turnaround fact.

a. 5 + 6 = (11) = __6__ + __5__

b. 3 + 4 = (7) = __4__ + __3__

c. 7 + 8 = (15) = __8__ + __7__

d. 4 + 5 = (9) = __5__ + __4__

Use Doubles   27

★ Answers will vary. This is one example.

Name: _____

1. Complete the matching number fact.

$\underline{6} + \underline{8} = \underline{14}$

 What double did you use to figure out the answer?

What other double could you use?

2. Write the double fact.
Draw two more dots then write the 'near double' fact.

a.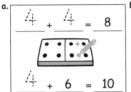

$\underline{4} + \underline{4} = 8$

$\underline{4} + \underline{6} = 10$

b.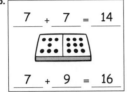

$7 + 7 = 14$

$7 + \underline{9} = 16$

3. Complete a fact to match each picture.

a.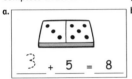

$\underline{3} + \underline{5} = \underline{8}$

b.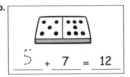

$\underline{5} + \underline{7} = \underline{12}$

Use Doubles

Name: _____

1. Write the double fact. Draw two more dots then write two 'near double' facts.

a.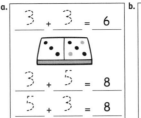

$\underline{3} + \underline{3} = 6$

$\underline{3} + \underline{5} = 8$

$\underline{5} + \underline{3} = 8$

b.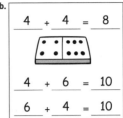

$4 + 4 = 8$

$4 + \underline{6} = 10$

$\underline{6} + \underline{4} = 10$

2. Complete each of these.

a.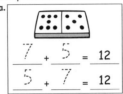

$\underline{7} + \underline{5} = 12$

$\underline{5} + \underline{7} = 12$

b.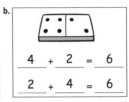

$\underline{4} + \underline{2} = 6$

$\underline{2} + \underline{4} = 6$

3. Complete these 'near double' facts.

a. $\underline{3} + 5 = 8$ **b.** $6 + \underline{8} = 14$ **c.** $4 + 6 = \underline{10}$

d. $\underline{2} + 4 = 6$ **e.** $7 + \underline{9} = 16$ **f.** $5 + 7 = \underline{12}$

4. Write the missing numbers.

a. $8 + 6 = \underline{14} = 6 + 8$ **b.** $10 + 8 = \underline{18} = 8 + 10$

Use Doubles

Name: _____

Jade had 9 girls and 6 boys at her party. How many guests in all?

1. a. These ten-frames show 9 counters.
Draw 6 more counters.

 Start here

b. Write the total. $\underline{15}$

 What did you notice?

c. Complete the sentence.

$9 + 6$ **is the same as** $10 + \underline{5}$

2. a. These ten-frames show 8.
Draw 4 more counters.

 Start here

b. Write the total. $\underline{12}$

c. Complete the sentence.

$8 + 4$ **is the same as** $10 + \underline{2}$

Make a Ten

Name: _____

1. Draw more counters then complete the sentence.

a.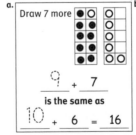

Draw 7 more

$\underline{9} + \underline{7}$

is the same as

$\underline{10} + \underline{6} = 16$

b.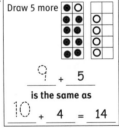

Draw 5 more

$\underline{9} + \underline{5}$

is the same as

$\underline{10} + \underline{4} = 14$

c.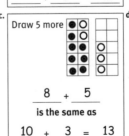

Draw 5 more

$\underline{8} + \underline{5}$

is the same as

$\underline{10} + \underline{3} = 13$

d.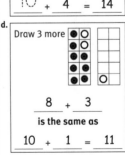

Draw 3 more

$\underline{8} + \underline{3}$

is the same as

$\underline{10} + \underline{1} = 11$

2. For each of these, draw an arrow to a number sentence below that has the same answer. Write the answer.

a. $9 + 8$ **b.** $8 + 6$ **c.** $9 + 3$

$10 + 2 = \underline{12}$ $10 + 7 = \underline{17}$ $10 + 4 = \underline{14}$

Make a Ten

Answers

Name: _____

Rosie bought three books.
They cost $7, $6, and $4.
What was the total cost?

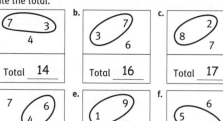

1. **a.** Write the numbers.

$7 + ($6 + $4)

Look for an easy way to figure out the answer.

b. Ring two numbers that add to make ten.

c. Write the total. $17

2. Suppose the books were $3, $5, and $7.

Find a pair of numbers that add nicely together.

Write the number sentence.

$3 + $7 + $5 = $15

Use Compatible Pairs

Name: _____

1. For each of these, ring the two numbers you would add first. Write the total.

a. 7 3 4 — Total 14
b. 7 3 6 — Total 16
c. 2 8 7 — Total 17
d. 7 6 4 — Total 17
e. 9 1 7 — Total 17
f. 6 5 5 — Total 16

2. Figure out the total score for each of these targets.

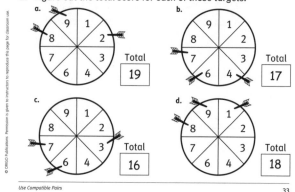

a. Total 19
b. Total 17
c. Total 16
d. Total 18

Use Compatible Pairs

Name: _____

1. Figure out these in your head. Write the answers.

a. 6 + 6 = 12   **b.** 30 + 30 = 60   **c.** 4 + 5 = 9

d. 6 + 8 = 14   **e.** 9 + 5 = 14   **f.** 7 + 5 + 3 = 15

2. Write the answer.

7 + 9 = 16

Explain two ways you could figure this out.

3. Write two near doubles you can solve using this double. ★

3 + 5 = 8 | 3 + 3 | 5 + 3 = 8

4. For each target, draw three 🖋 to show how you could make the total score.

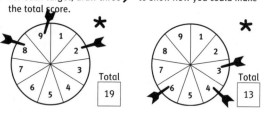

★ Total 19
★ Total 13

Check Up

Name: _____

Figure out each of these and write the answer.
Find the part below that has the answer and color it to match.

a. 9 + 9 = 18 (light blue)   **b.** 40 + 40 = 80 (red)

c. 7 + 8 = 15 (orange)   **d.** 3 + 5 = 8 (dark blue)

e. 20 + 20 = 40 (purple)   **f.** 9 + 5 = 14 (dark green)

g. 8 + 3 = 11 (light green)   **h.** 5 + 6 + 5 = 16 (yellow)

Just for Fun

★ Answers will vary. This is one example.

Answers

SUBTRACTION STRATEGIES

Count back

11 – 2   is   11 ... 10 ... 9.

45 – 3   is   45 ... 44 ... 43 ... 42.

Count on

8 – 6 = 2   because   6 + 2 = 8

Use doubles

18 – 9 = 9   because   9 + 9 = 18

WARM UP 13

Name: _____

There were 7 shells in the sand.
A wave washed 2 away.
How many are left?

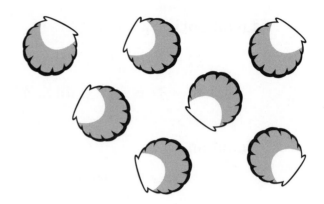

1. a. Figure out the answer.
        Cross out shells to help.

    b. Write the number fact.

    ┌─────────────────────────────────┐
    │   7   │
    │ ___ – ___ = ___   │
    └─────────────────────────────────┘

How do you know your answer is correct?

2. a. This picture shows 9 shells.
        Cross out 1 shell.
        How many are left?

    b. Write the number fact.

    ┌─────────────────────────────────┐
    │ ___ – ___ = ___   │
    └─────────────────────────────────┘

3. Write a fact to match this picture.

    ┌─────────────────────────────────┐
    │ ___ – ___ = ___   │
    └─────────────────────────────────┘

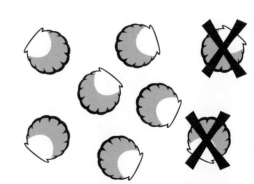

Count Back

1. Cross out shells. Complete the fact.

a.
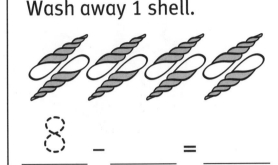
Wash away 1 shell.

8

___ – ___ = ___

b.
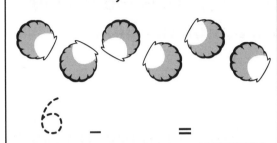
Wash away 2 shells.

6

___ – ___ = ___

c.
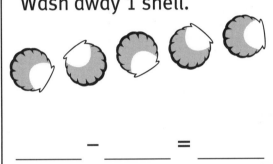
Wash away 1 shell.

___ – ___ = ___

d.
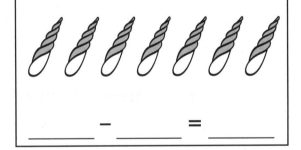
Wash away 1 shell.

___ – ___ = ___

2. Write a fact to match each picture.

a.
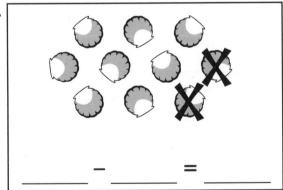

___ – ___ = ___

b.
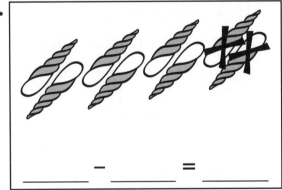

___ – ___ = ___

c.
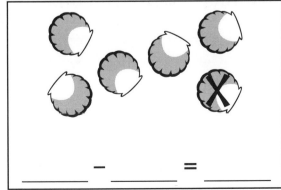

___ – ___ = ___

d.
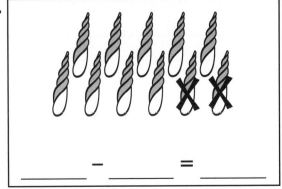

___ – ___ = ___

WARM UP 14

Name: _____

There are 12 children on a bus. Two children step off at the first stop. How many children are left on the bus?

1. Write a number fact to match the story.

   12 – _____ = _____

How did you figure out the answer?

2. Suppose 3 more children stepped off at the next stop.
   How many would be left on the bus?
   Write the number fact.

   _____ – _____ = _____

3. Make up stories to match these facts. Write the answers.

   _____ – 2 = _____   _____ – 1 = _____

Name: _____

1. Reduce each price by $1. Write the new price.

a. 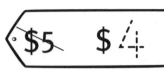   b. 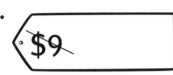   c.

2. Reduce these prices by $2.

a. 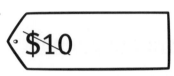   b.    c.

3. Reduce these prices by $3.

a.    b. 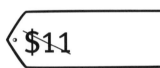   c. $6

4. By how much have these prices been reduced?

a. 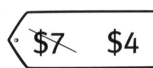   b. $7 $4   c. 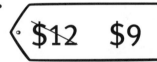 $12 $9

_____   _____   _____

5. Write some numbers to make each set true.

a.

11 − 1 = ___
11 − 2 = ___
11 − 3 = ___

b.

___ − 1 = ___
___ − 2 = ___
___ − 3 = ___

WaRM UP 15

Name: _____

A storekeeper has a pack of 12 glue sticks. She sells 3 sticks. How many are left?

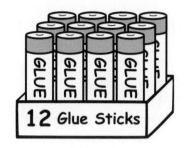

12 Glue Sticks

1. a. Figure out the answer in your head.
Cross out glue sticks to help.

   b. Write a number sentence to show what happened.

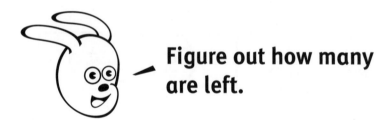

12 – _____ = _____

2. The storekeeper sold 2 scissors from this pack.

18

Figure out how many are left.

Write a number sentence to show what happened.

_____ – _____ = _____

How could you check your answer in your head?

Count Back

Name: _____

1. Sell 2 from each pack. How many are left? Write the answer.

a.

_____

b.
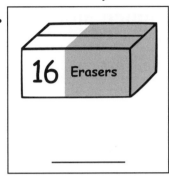
_____

c.

_____

2. Sell 3 from each pack. How many are left? Write the answer.

a.
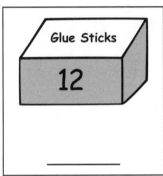
_____

b.

_____

c.

24 Rulers
_____

3. Write the answers.

a.

$17 - 1 =$ _____

$17 - 2 =$ _____

$17 - 3 =$ _____

b.

$26 - 1 =$ _____

$26 - 2 =$ _____

$26 - 3 =$ _____

c.

$39 - 1 =$ _____

$39 - 2 =$ _____

$39 - 3 =$ _____

d.

$15 - 1 =$ _____

$15 - 2 =$ _____

$15 - 3 =$ _____

e.

$25 - 1 =$ _____

$25 - 2 =$ _____

$25 - 3 =$ _____

f.

$34 - 1 =$ _____

$34 - 2 =$ _____

$34 - 3 =$ _____

WARM UP 16

Name: _____

Jade has saved $6. She needs $8 to buy a paint set. How much more money does she need?

1. **a.** Draw jumps to show how you could count on to 8 to figure out the answer.

| 1 | 2 | 3 | 4 | 5 | 6 | 7 | 8 | 9 |

b. Complete this sentence.

8 – 6 = _____ **because** 6 + _____ = 8

2. Suppose Jade had saved $5 and the paint set was $6.

a. Show how you could figure it out.

| 1 | 2 | 3 | 4 | 5 | 6 | 7 | 8 | 9 |

b. Complete this sentence.

6 – 5 = _____ **because** 5 + _____ = 6

Name: _____

1. Complete each of these. Use the track to help you.

1	2	3	4	5	6	7	8	9	10	11

a.

Spend $7

$8

8 − 7 = _____

because

7 + _____ = 8

b.

Spend $5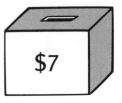

$7

7 − 5 = _____

because

5 + _____ = 7

c.

Spend $3

$4

4 − 3 = _____

because

3 + _____ = 4

d.

Spend $7

$9

9 − 7 = _____

because

7 + _____ = 9

e.

Spend $9

$11

11 − 9 = _____

because

9 + _____ = 11

f.

Spend $8

$10

10 − 8 = _____

because

8 + _____ = 10

WARM UP 17

Name: _____

**There are 8 ducks in all.
Four are in the pond.
How many are out of the pond?**

**Think of an addition fact that
will help you find the answer.**

1. Complete the
subtraction fact.

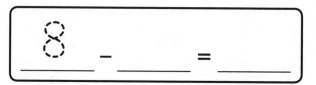

2. a. Write the missing number to make this balance.
Try using an addition fact to help.

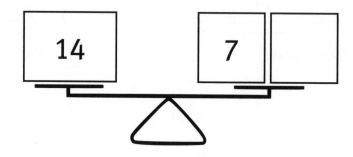

b. Write the addition fact
you used.

_____ + _____ = _____

c. Write the subtraction fact.

_____ − _____ = _____

1. Write the missing numbers to make these balance.
Complete two matching facts.

a.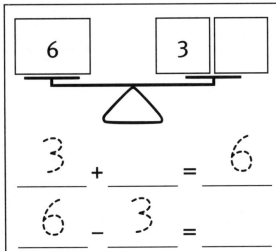

3 + _____ = 6

6 − 3 = _____

b.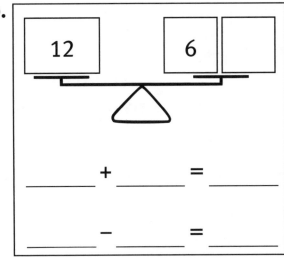

_____ + _____ = _____

_____ − _____ = _____

c.

| 18 | | 9 | |

_____ + _____ = _____

_____ − _____ = _____

d.

| 16 | | 8 | |

_____ + _____ = _____

_____ − _____ = _____

2. Complete each of these.

a.

5 + _____ = 10

so

10 − 5 = _____

b.

7 + _____ = 14

so

14 − 7 = _____

WARM UP 18

Name: _____

Ben has 40 cents. He needs 80 cents
to buy a drink. How much more
money does he need?

Use an addition fact
to help you find the answer.

1. Complete the
    subtraction sentence.

    80¢ _ 40¢ = _____
    ___ ___   ___

2. Beth needs 60 cents to buy an eraser.
    She only has 30 cents.

    Use addition to figure out
    how much more she needs.

    a. Write the addition sentence
        you used.   _____ + _____ = _____

    b. Write the subtraction sentence. _____ − _____ = _____

What other way could you figure out 60 − 30?

Use Doubles

Name: _____

1. Complete the addition sentence that will help you figure out the problem below. Write the answer.

a.
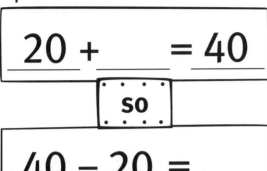
$$20 + \_\_\_ = 40$$
so
$$40 - 20 = \_\_\_$$

b.
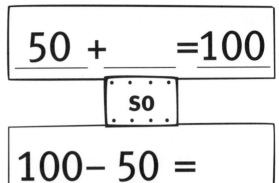
$$50 + \_\_\_ = 100$$
so
$$100 - 50 = \_\_\_$$

c.
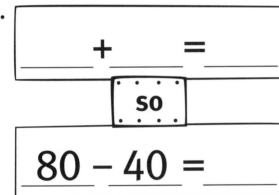
$$\_\_\_ + \_\_\_ = \_\_\_$$
so
$$80 - 40 = \_\_\_$$

d.
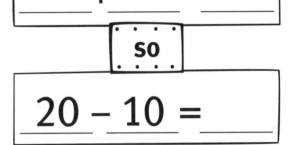
$$\_\_\_ + \_\_\_ = \_\_\_$$
so
$$20 - 10 = \_\_\_$$

2. Draw arrows to connect each spaceship to its correct planet.

WARM UP 19

Name: _____

**There were 50 children on camp.
If 25 children were girls,
how many were boys?**

Summer Camp

1. a. Write the subtraction
   sentence. Use a related
   double to help.

   50 − _____ = _____

   b. Write the double you used.   _____ + _____ = _____

2. Suppose 24 children were on camp and 12 children were
   girls. How many children were boys?

**Think of a related double
that will help you find the answer.**

Write the subtraction sentence.   _____ − _____ = _____

**Tell another way you could
figure out 24 − 12.**

Name: _____

1. Draw arrows to connect each subtraction sentence to a related double. Write the answers.

a.
$$70 - 35 = \text{\_\_\_\_\_}$$

b.
$$30 - 15 = \text{\_\_\_\_\_}$$

c.
$$90 - 45 = \text{\_\_\_\_\_}$$

d.
$$50 - 25 = \text{\_\_\_\_\_}$$

$$50 = 25 + 25$$

$$90 = 45 + 45$$

$$70 = 35 + 35$$

$$30 = 15 + 15$$

2. The answers to these subtraction sentences are on the kites. Write the matching sentence above each answer.

$$86 - 43 \qquad 42 - 21 \qquad 64 - 32$$

a.
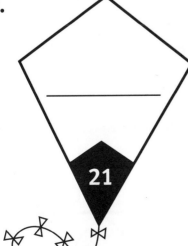
21

b.
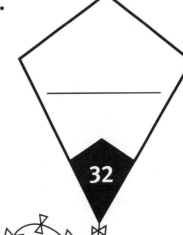
32

c.
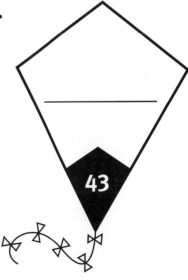
43

CHECK UP 3

Name: _____

1. Figure out these in your head. Write the answers.

   a. $9 - 2 =$ _____   **b.** $11 - 3 =$ _____   **c.** $16 - 2 =$ _____

   d. $10 - 8 =$ _____   **e.** $8 - 4 =$ _____   **f.** $40 - 20 =$ _____

2. **a.** Complete this number sentence.

   $60 - 30 =$ _____

   Use addition to help you.

   b. Write the addition sentence you used.   _____ + _____ = _____

3. Write the missing numbers to make these balance.

   a.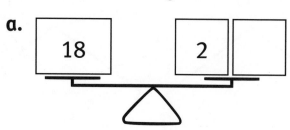

   18   2   [ ]

   b.

   11   9   [ ]

   c.

   16   8   [ ]

   d.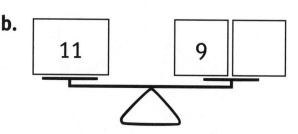

   24   12   [ ]

Name: _____

Figure out each of these and write the answer.
Find the part below that has the answer and color it to match.

a. $7 - 2 =$ _____ (light blue)   **b.** $12 - 9 =$ _____ (light green)

c. $56 - 3 =$ _____ (orange)   **d.** $8 - 7 =$ _____ (yellow)

e. $18 - 9 =$ _____ (purple)   **f.** $20 - 10 =$ _____ (dark green)

g. $100 - 50 =$ _____ (red)   **h.** $70 - 35 =$ _____ (dark blue)

WARM UP 13

Name: _____

There were 7 shells in the sand.
A wave washed 2 away.
How many are left?

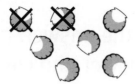

1. a. Figure out the answer.
        Cross out shells to help.

    b. Write the number fact.

    $\underline{7} - 2 = 5$

    How do you know your answer is correct?

2. a. This picture shows 9 shells.
        Cross out 1 shell.
        How many are left?

    b. Write the number fact.

    $9 - 1 = 8$

3. Write a fact to match this picture.

    $8 - 2 = 6$

WORK OUT 13

Name: _____

1. Cross out shells. Complete the fact.

    a. Wash away 1 shell.
        $8 - 1 = 7$

    b. Wash away 2 shells.
        $6 - 2 = 4$

    c. Wash away 1 shell.
        $5 - 1 = 4$

    d. Wash away 1 shell.
        $7 - 1 = 6$

2. Write a fact to match each picture.

    a. $10 - 2 = 8$

    b. $8 - 2 = 6$

    c. $6 - 1 = 5$

    d. $11 - 2 = 9$

WARM UP 14

Name: _____

There are 12 children on a bus.
Two children step off at the
first stop. How many children
are left on the bus?

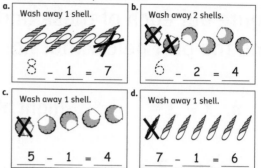

1. Write a number fact to match the story.

    $12 - 2 = 10$

    **How did you figure
    out the answer?**

2. Suppose 3 more children stepped off at the next stop.
    How many would be left on the bus?
    Write the number fact.

    $10 - 3 = 7$

3. Make up stories to match these facts. Write the answers.

    $\underline{7} - 2 = 5$ $\underline{8} - 1 = \underline{7}$ ★

WORK OUT 14

Name: _____

1. Reduce each price by $1. Write the new price.

    a. $5 $4   b. $9 $8   c. $6 $5

2. Reduce these prices by $2.

    a. $10 $8   b. $7 $5   c. $4 $2

3. Reduce these prices by $3.

    a. $9 $6   b. $11 $8   c. $6 $3

4. By how much have these prices been reduced?

    a. $11 $9   b. $7 $4   c. $12 $9
       $2 $3 $3

5. Write some numbers to make each set true.

    a.
    $11 - 1 = 10$
    $11 - 2 = 9$
    $11 - 3 = 8$

    b. ★
    $9 - 1 = 8$
    $9 - 2 = 7$
    $9 - 3 = 6$

★ **Answers will vary. This is one example.**

WARM UP 15

Name: _____

A storekeeper has a pack of 12 glue sticks. She sells 3 sticks. How many are left?

12 Glue Sticks

1. a. Figure out the answer in your head. Cross out glue sticks to help.

   b. Write a number sentence to show what happened.

   $12 - 3 = 9$

2. The storekeeper sold 2 scissors from this pack.

   → Figure out how many are left.

   Write a number sentence to show what happened.

   $18 - 2 = 16$

   How could you check your answer in your head?

48   Count Back

WORK OUT 15

Name: _____

1. Sell 2 from each pack. How many are left? Write the answer.

   a. 20 Crayons — 18
   b. 16 Erasers — 14
   c. 24 Note Pads — 22

2. Sell 3 from each pack. How many are left? Write the answer.

   a. Glue Sticks 12 — 9
   b. 15 Bulldog Clips — 12
   c. 24 Rulers — 21

3. Write the answers.

   a.
   $17 - 1 = 16$
   $17 - 2 = 15$
   $17 - 3 = 14$

   b.
   $26 - 1 = 25$
   $26 - 2 = 24$
   $26 - 3 = 23$

   c.
   $39 - 1 = 38$
   $39 - 2 = 37$
   $39 - 3 = 36$

   d.
   $15 - 1 = 14$
   $15 - 2 = 13$
   $15 - 3 = 12$

   e.
   $25 - 1 = 24$
   $25 - 2 = 23$
   $25 - 3 = 22$

   f.
   $34 - 1 = 33$
   $34 - 2 = 32$
   $34 - 3 = 31$

Count Back   49

WARM UP 16

Name: _____

Jade has saved $6. She needs $8 to buy a paint set. How much more money does she need?

1. a. Draw jumps to show how you could count on to 8 to figure out the answer. ★

   | 1 | 2 | 3 | 4 | 5 | 6 | 7 | 8 | 9 |

   b. Complete this sentence.

   $8 - 6 = 2$ because $6 + 2 = 8$

2. Suppose Jade had saved $5 and the paint set was $6.

   a. Show how you could figure it out. ★

   | 1 | 2 | 3 | 4 | 5 | 6 | 7 | 8 | 9 |

   b. Complete this sentence.

   $6 - 5 = 1$ because $5 + 1 = 6$

50   Count On

WORK OUT 16

Name: _____

1. Complete each of these. Use the track to help you.

   | 1 | 2 | 3 | 4 | 5 | 6 | 7 | 8 | 9 | 10 | 11 |

   a. Spend $7 $8
   $8 - 7 = 1$
   because
   $7 + 1 = 8$

   b. Spend $5 $7
   $7 - 5 = 2$
   because
   $5 + 2 = 7$

   c. Spend $3 $4
   $4 - 3 = 1$
   because
   $3 + 1 = 4$

   d. Spend $7 $9
   $9 - 7 = 2$
   because
   $7 + 2 = 9$

   e. Spend $9 $11
   $11 - 9 = 2$
   because
   $9 + 2 = 11$

   f. Spend $8 $10
   $10 - 8 = 2$
   because
   $8 + 2 = 10$

Count On   51

★ Answers will vary. This is one example.

Answers   61

WARM UP 17

Name: _____

There are 8 ducks in all.
Four are in the pond.
How many are out of the pond?

 Think of an addition fact that
will help you find the answer.

1. Complete the
   subtraction fact.

   $8 - 4 = 4$

2. a. Write the missing number to make this balance.
      Try using an addition fact to help.

   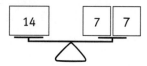

   14 | 7 | 7

   b. Write the addition fact
      you used.
      $7 + 7 = 14$

   c. Write the subtraction fact.
      $14 - 7 = 7$

52

Use Doubles

Name: _____

WORK OUT 17

1. Write the missing numbers to make these balance.
   Complete two matching facts.

   a.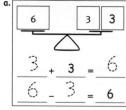
   6 | 3 | 3
   $3 + 3 = 6$
   $6 - 3 = 6$

   b.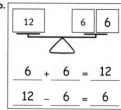
   12 | 6 | 6
   $6 + 6 = 12$
   $12 - 6 = 6$

   c.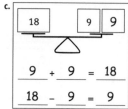
   18 | 9 | 9
   $9 + 9 = 18$
   $18 - 9 = 9$

   d.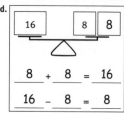
   16 | 8 | 8
   $8 + 8 = 16$
   $16 - 8 = 8$

2. Complete each of these.

   a.
   $5 + 5 = 10$
   so
   $10 - 5 = 5$

   b.
   $7 + 7 = 14$
   so
   $14 - 7 = 7$

Use Doubles

53

WARM UP 18

Name: _____

Ben has 40 cents. He needs 80 cents
to buy a drink. How much more
money does he need?

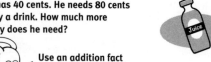

 Use an addition fact
to help you find the answer.

1. Complete the
   subtraction sentence.

   80¢ - 40¢ = 40¢

2. Beth needs 60 cents to buy an eraser.
   She only has 30 cents.

   Use addition to figure out
   how much more she needs.

   a. Write the addition sentence
      you used.
      30¢ + 30¢ = 60¢

   b. Write the subtraction sentence.
      60¢ - 30¢ = 30¢

    What other way could you figure out 60 - 30?

54

Use Doubles

Name: _____

WORK OUT 18

1. Complete the addition sentence that will help you figure out
   the problem below. Write the answer.

   a.
   $20 + 20 = 40$
   so
   $40 - 20 = 20$

   b.
   $50 + 50 = 100$
   so
   $100 - 50 = 50$

   c.
   $40 + 40 = 80$
   so
   $80 - 40 = 40$

   d.
   $10 + 10 = 20$
   so
   $20 - 10 = 10$

2. Draw arrows to connect each spaceship to its correct planet.

   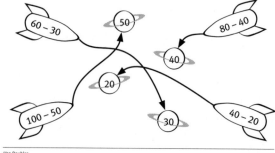

Use Doubles

55

WARM UP 19

Name: _____

There were 50 children on camp.
If 25 children were girls,
how many were boys?

Summer Camp

1. a. Write the subtraction
sentence. Use a related
double to help.

$$50 - 25 = 25$$

b. Write the double you used. $25 + 25 = 50$ ✱

2. Suppose 24 children were on camp and 12 children were
girls. How many children were boys?

Think of a related double
that will help you find the answer.

Write the subtraction sentence. $24 - 12 = 12$

Tell another way you could
figure out 24 – 12.

56   Use Doubles

Name: _____

WORK OUT 19

1. Draw arrows to connect each subtraction sentence to a
related double. Write the answers.

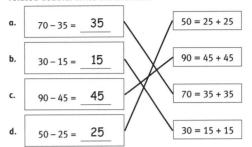

a. $70 - 35 = 35$ $50 = 25 + 25$

b. $30 - 15 = 15$ $90 = 45 + 45$

c. $90 - 45 = 45$ $70 = 35 + 35$

d. $50 - 25 = 25$ $30 = 15 + 15$

2. The answers to these subtraction sentences are on the kites.
Write the matching sentence above each answer.

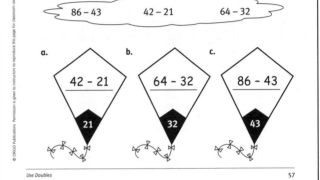

86 – 43   42 – 21   64 – 32

a.   b.   c.

42 – 21   64 – 32   86 – 43

21   32   43

Use Doubles   57

CHECK UP 3

Name: _____

1. Figure out these in your head. Write the answers.

a. $9 - 2 = 7$ b. $11 - 3 = 8$ c. $16 - 2 = 14$

d. $10 - 8 = 2$ e. $8 - 4 = 4$ f. $40 - 20 = 20$

2. a. Complete this number sentence.

$$60 - 30 = 30$$

Use addition
to help you.

✱

b. Write the addition sentence
you used. $30 + 30 = 60$

3. Write the missing numbers to make these balance.

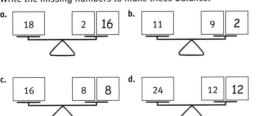

a. 18   2 16   b. 11   9 2

c. 16   8 8   d. 24   12 12

58   Check Up

Name: _____

JUST FOR FUN 3

Figure out each of these and write the answer.
Find the part below that has the answer and color it to match.

a. $7 - 2 = 5$ (light blue)   b. $12 - 9 = 3$ (light green)

c. $56 - 3 = 53$ (orange)   d. $8 - 7 = 1$ (yellow)

e. $18 - 9 = 9$ (purple)   f. $20 - 10 = 10$ (dark green)

g. $100 - 50 = 50$ (red)   h. $70 - 35 = 35$ (dark blue)

Just for Fun   59

✱ Answers will vary. This is one example.

Answers   63

Use this chart to help you add and subtract.

1	2	3	4	5	6	7	8	9	10
11	12	13	14	15	16	17	18	19	20
21	22	23	24	25	26	27	28	29	30
31	32	33	34	35	36	37	38	39	40
41	42	43	44	45	46	47	48	49	50
51	52	53	54	55	56	57	58	59	60
61	62	63	64	65	66	67	68	69	70
71	72	73	74	75	76	77	78	79	80
81	82	83	84	85	86	87	88	89	90
91	92	93	94	95	96	97	98	99	100

Have the children draw arrows to show their thinking.